Do-it-Yourself Solar-Powered Go-Kart

Simple DIY Solar Powered Go-kart Picture Guide for a Fun Weekend Project or Science Fair Project

By Eric Layton

Copyright 2017 by Eric Layton

Version 1.1

All rights reserved. No part of this book may be reproduced, stored in a retrieval system, or transmitted in any form or by means without prior permission from Eric Layton, except by a reviewer who may quote brief passages in a review to be printed in a newspaper, magazine, or journal.

DIY Solar-Powered Go-Kart

Figure 1. Eric Layton and his first solar powered go-kart

Acknowledgements

The building of this solar-powered go-kart would not have been possible without the contribution of several people and institutions. I would like to thank my engineering teacher, who assisted me in all the steps for the construction of this go-kart. I also benefited from resources provided by my high school.

I also appreciate the support and guidance provided by my dad, from whom I learned much about electric systems, and who provided me space in his garage for the execution of this project. He also taught me the most about electric vehicles, electric motors, etc. throughout my life. I appreciated all of his help with this project.

Finally, this book would not have been possible without the support of my wife, Flavia Leite, who edited the book and motivated me to get it published more than 12 years after I built the electric go-kart.

Preface

In 2004, during my junior year of high school, my teacher proposed that I created an electric car for a science fair project. I had originally proposed building a hovercraft, but the cost of this project and depth of building a hovercraft was not known at that time. This idea however proved unfeasible given that my budget for this project was only $3,000.

In this book, I explain my experience in building a solar-powered go-kart. I narrate each step of the process so the reader can follow what I did. I do have a YouTube channel located at www.youtube.com/nuclearboy2003 that has some videos of the construction and operation of the DIY electric go-kart. I highly recommend watching these videos as they will give you details not shown in this video.

I hope that you will find this book helpful to guide you on how to construct a solar powered go-kart.

Introduction

Go-karts, as with any other vehicle, can be powered by gasoline, natural gas, or electricity. Go-karts are usually powered by gasoline. The gasoline provides a powerful source of energy for the go-kart motor. However, gasoline has the disadvantage of being a non-renewable energy source that emits unhealthy and unneeded pollution.

Powering a go-kart with natural gas is impractical, because natural gas vehicles must have space for holding a large gas tank. Go-karts are small, so they have no space for holding a natural gas tank. Also, many people would find it uncomfortable to have a large holding tank of flammable gas in a vehicle designed to go off-road, possibly flip, bounce frequently, etc.

Powering go-karts with solar energy has advantage. The sun is a renewable source of energy that does not deplete as its energy is used. Moreover, it is a non-polluting source of energy since its use does not release carbon dioxide or other pollutants.

The Original Idea

My initial idea was to create a solar-powered electric vehicle from an existing car. I thought it would be easier to purchase an already-built vehicle, such as a dune-buggy, and transform it into solar-powered electric car. After all, most electric vehicles at that time were gasoline or diesel vehicles that had been converted to electric cars through electric vehicle conversion kits.

To create a solar-powered electric vehicle, I needed to buy a car suitable to the project. Given my budget limitations, it could not be too expensive, and it had to be in good condition. I set out to check Craigslist postings to see what was available for purchase. Craigslist was gaining popularity in 2004, and as of December 2017, is still a great place to find used go-karts for less than $500. I tried a quick search in several cities in Florida on Craigslist (Miami, Orlando, Tampa) and found over 50 go-kart frames (some even with gasoline engines) for less than $500 in December 2017.

From the first vehicle that I examined in 2004, I found it difficult to find an optimum car for conversion. One of the first vehicles that I examined was an old Volkswagen dune buggy that was for sale on Craigslist for $2,000. Below are pictures of this vehicle:

Figure 2. Blue dune buggy for sale on Craigslist

My budget for this project was $3,000. Purchasing a $2,000 dune buggy would consume 2/3 of the budget. The owner of the blue dune buggy was not willing to negotiate on the price, and the damage to the vehicle was extensive.

After discussing this project with some local engineers, including my engineering teacher, I decided to find a cheaper alternative. However, had this vehicle been in good condition, it would have been a great choice for a gas-to-electric conversion.

Preliminary sketches of the go-kart

Having not found a good vehicle for purchase, I opted to use a go-kart for the gas-to-solar conversion. I made a few rough sketches of a possible design for the go-kart. As a student in 2004 and 2005, I often made my designs on legal pads since the school provided these to the students. Figure 3 shows a general sketch of a go-kart.

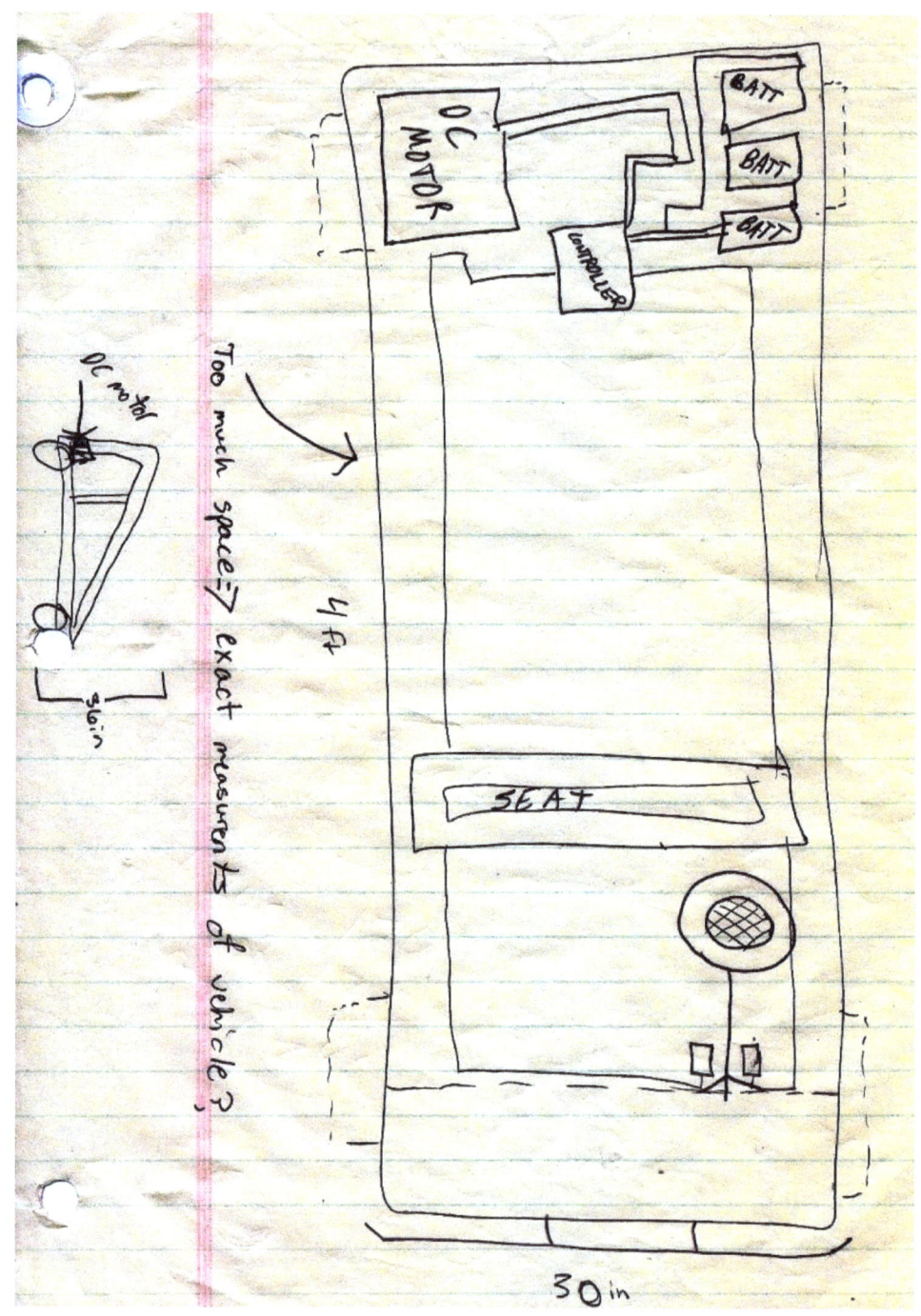

Figure 3. First sketch of the solar-powered go-kart with the motor, batteries, controller, and frame represented

Figure 4 shows a rough sketch of how the electric circuit would be connected to the solar panels. The letter "A" that is circled on the sketch in Figure 4 represents an ammeter, which is a meter that is used to measure amperage across the circuit. The letter "V" that is circled in several locations on the sketch (also on Figure 4) represents a voltmeter, which is a meter that is used to measure voltage across a circuit.

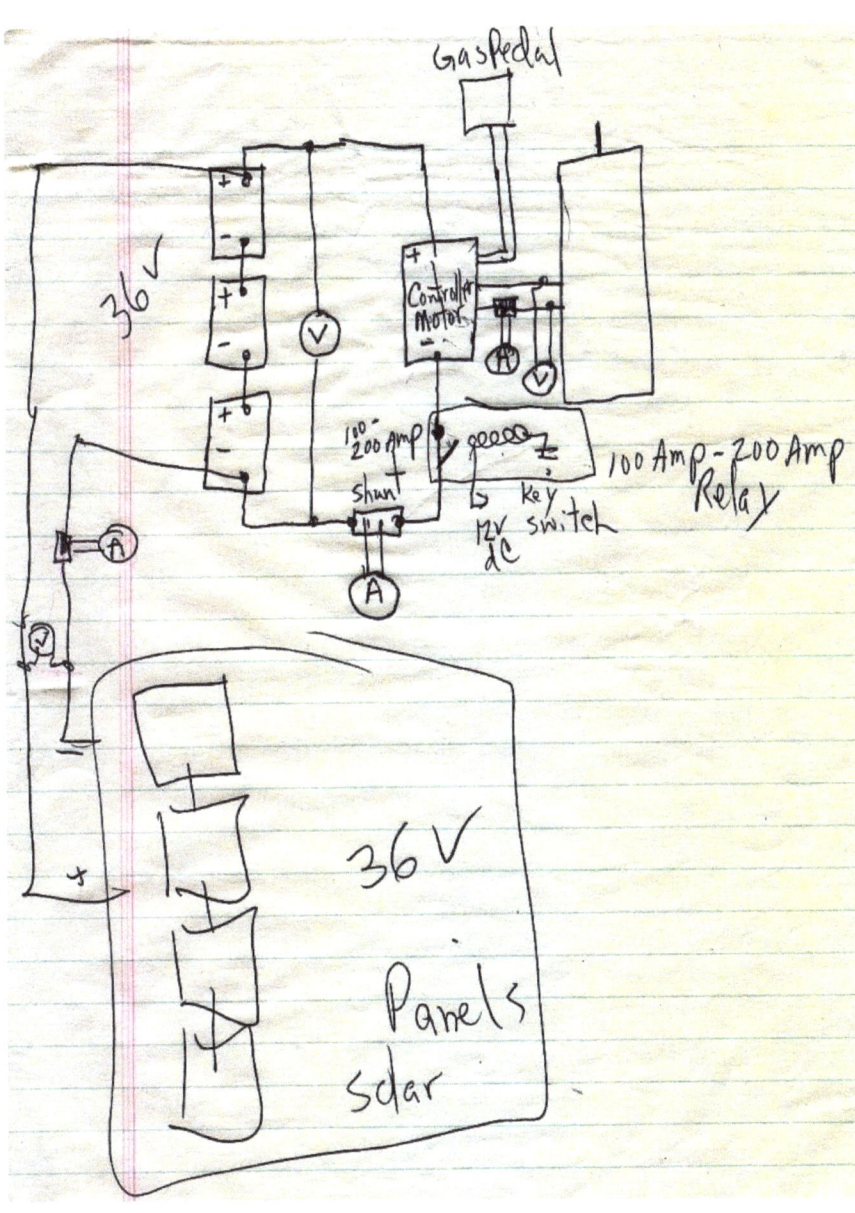

Figure 4. First sketch of the electronic system with the solar panels, relays, and batteries represented

Purchasing the go-kart kit

I purchased a go-kart kit for $500 from the website of Northern Tool (www.NorthernTool.com). This kit had everything I needed to construct the go-kart, except for the engine, the clutch, and a roll cage. The go-kart kit included wheel halves, bucket seat, seat cover, chain, steering shaft, main frame, wheels, pedals, the brake linkage to the rear wheels, and an instruction guide on how to assemble the parts. Below is the actual posting from Northern Tool's website:

Figure 5. The Northern Tools Go-kart Kit that was purchased for this project

This kit was a great purchase considering that my budget was $3,000. There were no already-assembled go-kart frames for less than $1,000 that could be found in good condition or new on Craigslist or anywhere else on the internet. This go-kart frame was available from several retailers, but the cheapest option with shipping included was from Northern Tools. The total cost of purchasing this go-kart kit with UPS shipping came to $560.58.

Existing go-karts are easier to retrofit with solar panels and an electric motor because they just require removing the gasoline engine and installing the solar panels and the electric motor. Therefore, if you find a go-kart on Craigslist for less than $500 to $600, I highly recommend you inspect it to see if it is in good condition (no rust, brakes work, roll bar assembly is good, etc.) for purchase.

A go-kart kit like the one I purchased requires many hours to build. It took me about four hours to put the go-kart frame together. The positive side of buying the kit was that it was fun to work on building the frame myself. Figures 5 through 7 show details of the go-kart kit I bought.

Figure 6. Parts of the go-kart kit I purchased

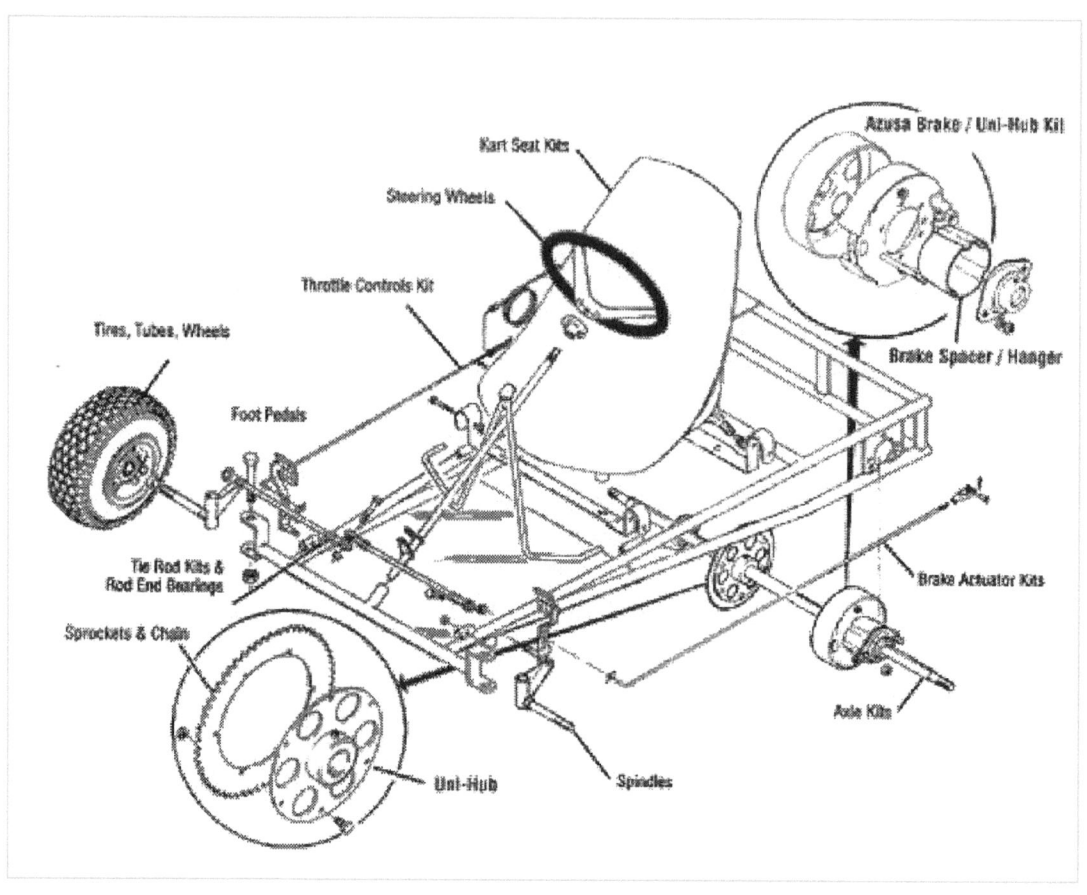

Figure 7. Image from kit manual showing how to assemble the go-kart that came with the kit

Designing a mounting structure for the solar panels

One key step in building a solar powered go-kart is to build a mounting type of structure for the solar panels over the base frame. This structure serves two purposes: it provides the structure over which the solar panels will be placed and it provides security/limited protection to the driver in case the go-kart rolls or tips over. Since my goal was to test whether the solar panels worked to charge the batteries and I had a limited budget, I used PVC to build a temporary structure. While using PVC for the mounting structure was a reasonable choice for my project, PVC is very flimsy and should NOT be used for protection. As a recommendation, you should NEVER use PVC as a roll cage – only as a temporary structure for the solar panels. My recommendation is to use metal for the roll cage and mount the solar panels on top of the roll cage. By using metal for the roll cage, the driver will be better protected if the car rolls over. At the time that I built this DIY electric go-kart, I did not possess the skills for welding or the budget for buying metal or having access to a welder.

Buying an electric motor kit

After about a month of research, I purchased an electric motor kit from a company called Cloud Electric based on extensive Google searches. Keep in mind I did this search on the internet back in 2004, so there are many more electric motor kits available. After a quick Google search in December 2017, for example, I found more than 10 comparable kits, including kits with a "pancake-style" electric motor. This type of motor is called a "pancake-style" because it looks like an extruded circle instead of a thin cylinder shape. I specifically chose a pancake-style electric motor due to the limited room for mounting the motor on the go-kart.

The motor kit included the following items:

1. Etek 8 Horsepower (HP) Continuous Motor (pancake-style)
2. Alltrax Programmable 24-48 Volt 300 Amp Controller
3. Fingertip Controller Pot (this is the lever or potentiometer used as the "gas")
4. 250 Amp Fuse
5. 500 Amp Keyed Cutoff Switch
6. 6' #4 Cable Red Wire
7. 6' #4 Cable Black Wire
8. (12) Wire Lugs

The kit was called the Cloud EV Electrathon Drive Kit and it cost $780, an excellent price at the time. I had budgeted $1,200 for the motor kit, so buying this kit helped me save money and my budget. Shipping for the kit was $85. The whole kit weighed 46 pounds. The weight of the kit was one of the deciding factors for the purchase. Kits that weigh less are better because they do not make the go-kart heavy. Weight can critically reduce the efficiency of the motor and battery life. I also considered the quality of the customer service in my decision to buy this kit. The owner of the website that sold me this kit, Steve Cloud, had quickly e-mailed me back with answers to my questions. Because of this personalized and fast reply, I knew that I would have immediate access customer service if I had any problems with the kit.

Figure 8 is a scan of a printout of the Cloud EV Electrathon Drive Kit from the Cloud Electric website:

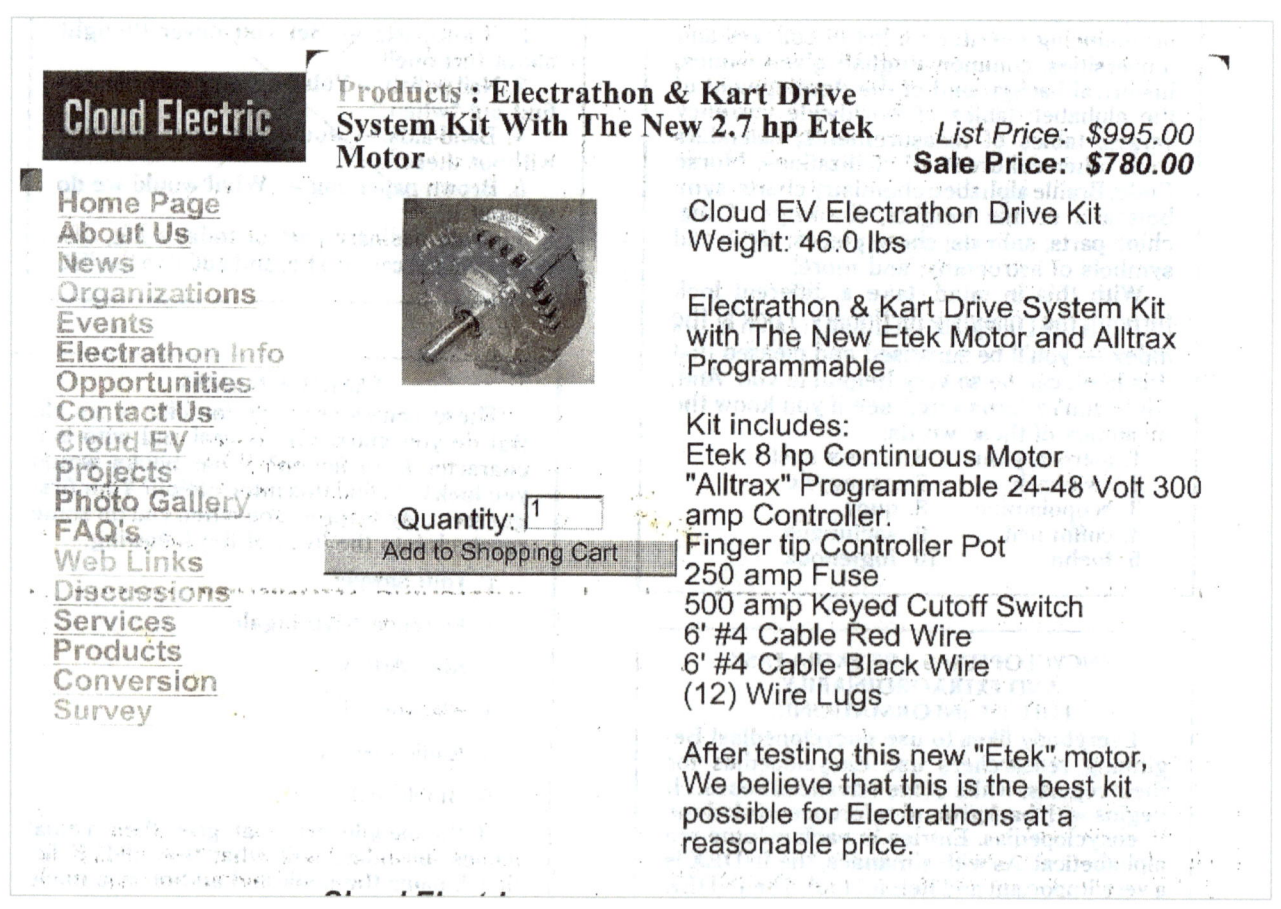

Figure 8. Cloud EV Electrathon Drive Kit

The Etek 8 HP electric motor was the best purchase of this kit due to the high quality of the motor manufactured by Briggs & Stratton. This motor is perfect for a go-kart because it does not require a high voltage to operate. The motor has 8 horsepower and functions within the ideal voltage of 24 to 36 volts. For a go-kart, this motor is very powerful (high torque). If I purchased only the motor, the Etek "pancake" electric motor would cost about $500. This 22.3-pound motor was almost half the weight of comparable electric motors for go-karts. A motor with lower weight is a plus for building go-karts. You do not want to buy a heavy motor, so the extra weight does not interfere with the kart's performance.

Below are two images of the Cloud Electric website showing the Etek 12-48 volt 8 HP DC motor (Figure 9 and Figure 10):

Products : * Permanent Magnet DC Motors
* : **Motor, Etek 12 - 48 volt 8 HP DC** Briggs & Straton - Etek
Weight: 22.3 lbs
List Price: $495.00
Sale Price: $425.00

Etek Motor Drive Sprockets

Etek Motor Information and Specifications

Note: This motor is no longer in Production (big mistake!). This is NOT the motor as described on the Briggs & Straton Web Site! Their new brushless motor will not be recommended by us and we have no plans to offer it for sale.

Etek Permanent Magnet "Pancake" Design, 20% lighter than standard series wound motors.

20 HP Peak, 8 HP continuous
Voltage........................50 VDC
Max Nonload Current.....6 AMP
Max Nonload Speed........3600 RPM
MIN Nonload Speed........3300 RPM
Min Speed at 160 LbIn.....3200 RPM
Max Current at 160 LbIn...150 AMPS
Shaft Size ....7/8"
Voltage Constant: 72 RPM per volt
Torque Constant: 1.14 in lbs/Amp
Continuous Current: 300 A 30 Sec.
Weight: 20.8 lbs
Motor Diameter: 7.91"
Motor Length: 5.64"
Shaft Diameter: 7/8"-3/16" Keyway

Quantity: 1
Add to Shopping Cart

FEATURES BENEFITS
50% smaller and 20% lighter (only 22.3 lbs.) than a competitive electric motor - high power to weight ratio.
Lightweight aluminum frame. Less turf compaction.
Provides DC (battery) electric power. Quiet, reliable power source.
Provides a maximum of 20 HP, 8 HP continuous. High efficiency extends run time.
"What makes the Etek motor technology unique is the use of copper bus bars rather than steel and copper wire as the basic building block of the

Figure 9. Specifications of the Etek 8 HP Motor (page 1)

armature. These copper bus bars are stamped, bent, coated and assembled into a thin rotary disk. End clips connect the tips of the bus bars to shorten the air gap between the magnets. Steel is inserted between the bus bars to shorten the air gap between the magnets. One of the most unique characteristics of the motor is that simply machining the edges of the copper bars produces the commutator. Since commutator is built in, there is no need for a separate assembly. The motor uses neodymium magnet technology.

The resulting motor construction is what conventional technologists would describe as a wave wound axial air gap brushed DC motor. The performance, however, is far from conventional. When compared to an equivalent wire wound DC motor capable of producing similar torque, several advantages are obvious. The first is that the Etek motor uses one-tenth the steel and one-half the copper as is used in a conventional DC motor. The second advantage is that the Briggs & Stratton motor is one-half the size of a competitive motor. While many designers today think of the motor and controller late in the design cycle, Briggs & Stratton believes that some designers may want to start with the company's motor size and design around it.

Possibly one of the most significant performance features of the Etek motor is efficiency. While demonstrating high efficiencies under no or small loads, conventional DC motors experience a dramatic drop in efficiency (as low as 70 percent are not uncommon) as they are loaded down. This simply means that nearly a third of the valuable energy in the on-board batteries is going toward producing heat and other unwanted power rather than motion. The Etek motor technology does not use battery energy to produce its electric field in the stator. This is done by the neodymium permanent magnets. The use of copper bus bars gives the motor an extremely low internal resistance, which results in greatly reduced losses."

Figure 10. Specifications of the Etek 8 HP motor (page 2)

Figure 11 shows the large fuse that came with the Cloud Electric Vehicle Kit go-kart kit. This 250-amp fuse was used to provide protection for the controller and the motor. Having a fuse is an absolute requirement in building an electric go-kart.

Figure 11. Amp fuse of the Electrathon Drive Kit (250 amp)

Figure 12 shows the front view of the Etek 8 HP electric motor. The Etek electric motor is a very efficient electric motor for a go-kart. It has a "pancake" motor configuration, which is different from the more common "elongated" motor configuration. This motor is thin, which makes it perfect for mounting on the go-kart frame. The two gold screws in the motor are the positive (+) and negative (-) terminals. The motor was quite heavy for lifting as it weighed about 21 pounds.

Figure 12. Front view of the Etek 8 HP electric motor

Figures 13 and 14 show the other side of the Etek 8 HP electric motor"

Figure 13. The other side of the Etek 8 HP style electric motor

Figure 14. Another view of the Etek 8 HP style electric motor

The controller

The controller is a critical component of the electric vehicle. It allows the correct voltage to be transferred from the batteries to the motor. When the pedal (this is called the "gas pedal" on a gasoline powered go-kart) is pressed, the pedal pulls on a device called a potentiometer, which is connected to the controller. Depending on the distance the potentiometer is pulled, which corresponds to the distance the driver depresses the pedal, the controller will enable the correct voltage and amperage to be drawn from the batteries and sent to the motor. This voltage/amperage draw is programmable in the controller. The user can program the controller to be more or less sensitive to the potentiometer. This is why the controller is often nicknamed the "brain" of the electric vehicle. The AXE controller that came in the kit (Figure 15) used a cable to connect to a PC for programming.

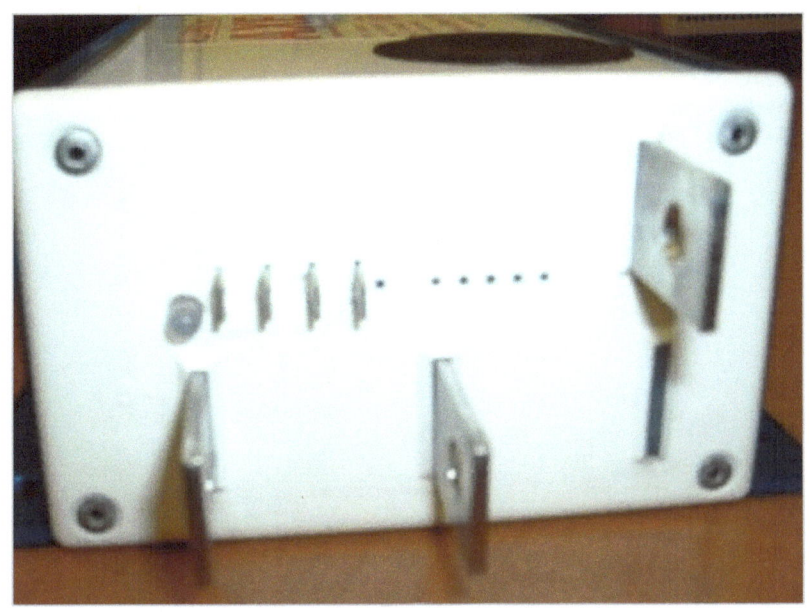

Figure 15. AXE controller of the electric go-kart

I decided to conduct a test to understand the importance of using a controller. In Figure 16, I connected the batteries directly to the electric motor without using a controller. A 12-volt battery was therefore directly connected to the Etek electric motor. This caused the motor to turn on and spin so fast that the Etek motor flipped over. With no type of controller or potentiometer to manage the voltage that goes to the electric motor, the motor will spin at full RPM (revolutions per minute) on the first turn of the shaft. However, when the controller is connected, a potentiometer can be used to control how fast the electric motor spins, hence determining how fast the go-kart will move forward.

Figure 16. Electric motor connected directly to a 12-volt battery (without the controller)

The motor kit came with a schematic that shows how the controller should be connected to the motor, the fuse, the potentiometer, and the batteries in the electric system (Figure 17). This schematic is not adequate for a solar-powered go-kart because it does not show how the solar panels will be connected to the electric system.

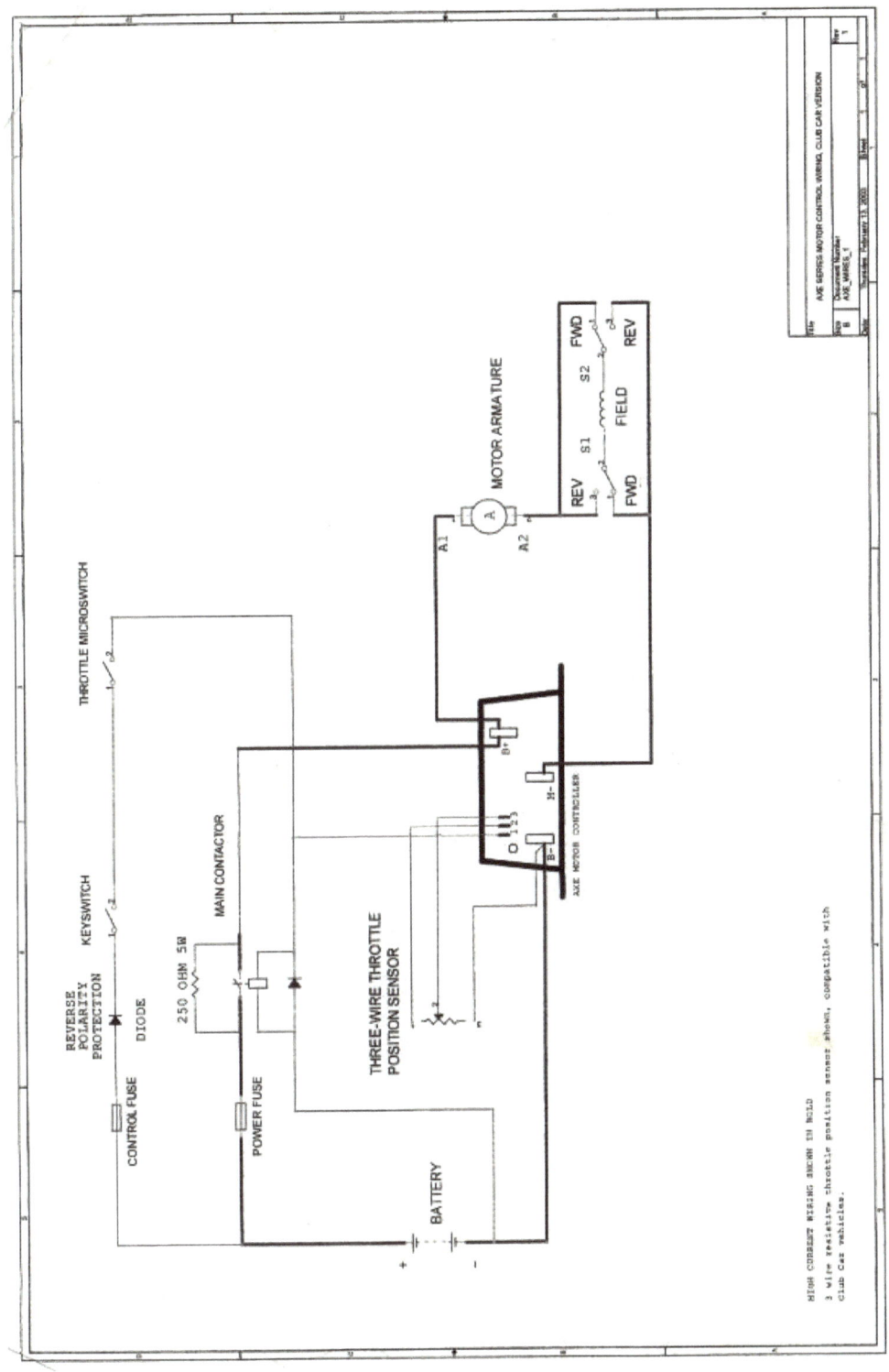

Figure 17. Alltrax controller schematic that came with the motor kit

I created a new schematic to show how the controller and the solar panels would be connected to the electric system for building a solar-powered go-kart (Figure 18). The different "V" circles represent where voltmeters are to be placed, and the "A" with the circle represents where an ammeter is to be placed.

Anytime amperage is measured, a permanently mounted device called a "shunt" must be used. This is designed to be built in series with the circuit, as shown in Figure 18. I purchased the shunt at Radio Shack, but it can easily be purchased over the internet.

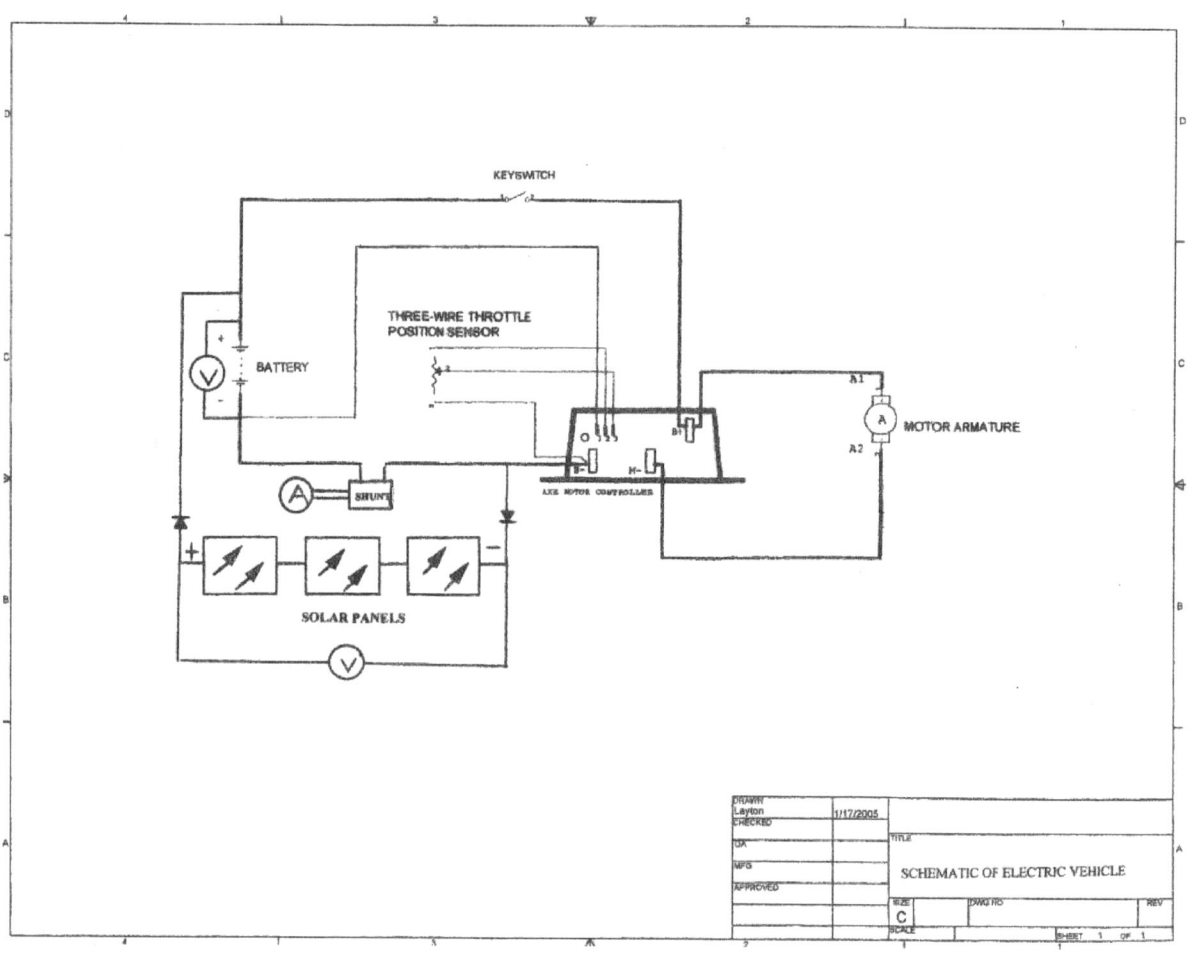

Figure 18. Modified circuit schematic showing how the solar panels and the controller are connected to the electric system

Obtaining and testing the batteries

The batteries I used for this project were 12-volt lead acid batteries. They were donated by a local company that did not need them anymore. They were used batteries from a UPS (Uninterruptible Power Source) battery backup for a large server farm of computers. These UPS batteries were being used to provide energy for servers and large computer systems in the case of a loss of power. In such situations, these emergency batteries would automatically supply the necessary power for several minutes until the main power supply returned.

I tested each battery to see if they were holding the charge. I used a battery checker from Harbor Freight Tools that had cost about $20. If you do not have a battery checker, another method for checking if a battery can hold charge and is "good" for an electric go-kart is to charge the battery overnight and then check the voltage with a multimeter the next day. A fully charged 12-volt battery should read roughly 13 to 14 volts fully charged. After testing the donated batteries, I found about 2 to 3 of them were "bad" by the battery tester and were properly recycled.

Building the go-kart frame

a. Painting the Frame

The original go-kart frame was a dull gray color when purchased from Northern Tools. I decided to the paint the frame red to indicate that this was a prototype, and because red was my favorite color at the time. I spray painted, sanded, and then applied a second coat of spray paint to the entire frame.

The entire frame was spray-painted with a grey base coat first, then several coats of outdoor red spray paint designed for metal surfaces that can easily bought from a hardware store. If you decide to paint the go-kart frame, it is necessary to do this painting well before any parts are installed. If you have a pre-made or already existing go-kart frame, then all parts need to be removed so that not everything is spray painted (covering the wheels, removing steering wheel, etc. before painting).

b. Assembling the frame

After painting the frame, I followed all the instructions that came with the go-kart kit to install all the parts, including the wheels, bearings, pedals, steering wheel, and seat. The entire process of assembling the basic components on the go-kart took about 4 to 5 hours.

After spray painting, the frame can be assembled. By following directions that came with the go-kart frame kit, the seat, steering wheel column/assembly, tie rods, pedals and wheels can be installed. This process takes about 2 hours to fit all components properly and was easily accomplished with focus as well as precision in mounting.

c. Building a wooden structure to hold the batteries

Wood was mounted to the back of the frame using U-shaped bolts purchased at the local hardware store. The wooden structure was a simple design to secure the batteries and the sides of the structure were only a few inches high. This structure was installed at the back of the go-kart. Several pictures can be found later in this book showing the back wooden structure of the go-kart.

d. Installing the batteries in the back of the go-kart

I placed several of the UPS batteries on the top of the wood structure in the back of the vehicle. Each of the batteries weighs about 10 pounds, so the number of batteries seen in Figure 19 and Figure 20 was the limit of the number of batteries that could fit on the back of the empty go-kart. Adding more batteries would cause a weight imbalance where the front end would completely lift in the air due to 50 pounds or more added to the back of the go-kart.

Six batteries were installed on the wooden shelf on the back of the go-kart. Since it was decided that the voltage of the system would be 36 volts, the batteries would need to be in a parallel circuit of 36 volts to match the voltage of the Etek electric motor. Each UPS battery was 12 volts, meaning that three of these batteries would need to be in a series circuit to add to 36 volts. In Figure 21 six batteries were installed on the wooden shelf so that there were two sets of three batteries. The two sets of 36 volts each were installed in parallel with each other since in a parallel circuit the voltage remains the same.

The weight of the batteries must be evenly distributed around the entire frame. On this electric go-kart outlined in this book, several locations had 2" x 4" wood and other pieces of wood to further extend the frame. This influences the center of gravity, especially with the heavy batteries and electric motor. Thus, the placement of batteries is critical and some experimentation will be required by the builder (without the driver sitting in go-kart) by placing the batteries around the frame.

Figures 19, 20, 21, and 22 show the assembled go-kart frame with the batteries placed on the custom-built wooden structure.

Figure 19. Go-kart mounted with all parts that came in the kit and wood board for placing the batteries

Figure 20. Mounting structure for the Etek electric motor was installed

Figure 21. The first set of batteries were installed in the rear of the go-kart

Figure 22. Rear view of the batteries and the sprocket/chain assembly

The go-kart was not very balanced with the heavy load in the back. The batteries caused the go-kart's front end to tilt up from the heavy weight in the rear end. The front wheels were starting to lift off the ground when not holding the go-kart down by force. The distribution of weight was all in the back of the go-kart without the batteries and driver in the go-kart. Tipping the go-kart backwards (the front wheels lifting off the ground) was a major issue in the beginning phases of building as well as deciding where to install the additional batteries.

Figure 23. Left side of the go-kart after wooden battery holder was constructed

e. Adding additional batteries between the feet of the driver

To avoid a weight imbalance and increase the energy of the go-kart, I added additional batteries underneath the steering column, between the feet of the driver. The bottom of the go-kart where the driver put their feet had space directly below the steering column. After more consideration, it was decided to mount the UPS batteries directly underneath this steering column. The driver's feet go on each side of the go-kart and can rest easily on the pedals. Once the batteries were installed on the red pan underneath of the driver's column, the center of gravity moved forward underneath the driver's seat.

f. Installing the Motor

Whether you choose a pancake motor or an elongated shaped electric motor, mounting the electric motor to the frame will be different for every motor and go-kart frame. L-shaped brackets were used to mount the motor to the wooden board that I mounted the electric motor. These brackets helped stabilize the electric motor. A chain was used since this came with the go-kart kit.

On the right-hand side of Figure 24 is the Etek electric motor. After the go-kart was completed, a bracket mount was made for the black right-hand side of the electric motor to help stabilize the vibration. Occasionally while testing, the chain would come off because of the vibration of the motor. Both sides of the motor needed to be stabilized and secured so it would not move to due to vibration. Heavier brackets were used to mount the electric motor, which can partly be seen in Figure 26.

Figure 24. Rear view of the placement of the electric motor

g. Installing the Controller

Figure 25 has the AXE controller mounted, which was drilled and mounted to the side of the batteries. Figure 26 shows more detail with the AXE united mounted. The black and gold unit to the front of the controller in Figure 26 is called a "shunt", which is the connector that is used to measure amperage across a circuit.

Figure 25. AXE controller screwed into the wood

Figure 26. Battery holder, AXE controller, and shunt installed

Figure 27. Side view of battery holder, controller, and motor mount

h. Installing the shunt and voltmeter

A shunt must be installed in parallel in a circuit and never in series with other electrical components. A shunt is installed opposite a voltmeter because a voltmeter is installed in series with a circuit. In Figure 4 at the beginning of the book, I show in the drawing the letter "A" that is circled and it is drawn in parallel with each individual circuit while a voltmeter is designated with the letter "V". The voltmeter symbols on this drawing in Figure 4 are all installed in series with a circuit. Keep in mind that an ammeter and a voltmeter can be installed on the same circuit.

I installed a total of 3 shunts and 3 voltmeters on the go-kart frame. The displays that were used were cheap multimeters from a store called Harbor Freight that cost a few dollars each. These multimeters were then installed on the dashboard of the electric go-kart.

These were the locations of where the shunts and voltmeters were installed:
1. In the circuit with the solar panels to show the amperage/voltage (and therefore power) being charged from the solar panels
2. In the circuit from the controller to the electric motor to show how much power the motor was using
3. In the circuit with the batteries to show the overall voltage of the battery system (the voltage would show me how charged the batteries are)

i. Installing the sprockets and chain assembly

The sprocket, which is the main gear that connects to the axle, was easy to install as the sprocket simply slides on the axle. One of the most difficult parts of the assembly of the go-kart was connecting the chain to the electric motor sprocket and the main sprocket on the axle. This process required using L-shaped brackets and heavy bolts to create a sturdy mount for the electric motor. Perfect tension was required so that the chain did not slip off either sprocket. Figure 28 shows the L brackets and the tension required for the go-kart to properly function. Early in my test trials I found that I had not created enough tension with the chain, so the chain would constantly slip off the sprockets.

Also, I did not buy any additional sprockets and only used the sprockets that came with the Etek electric motor and the go-kart kit. When you glance at Figure 29 and notice that I have an inefficient gear ratio because of the small sprocket on the motor to the larger sprocket on the axle. My "transmission" for this go-kart means that I would not go as fast as I had initially predicted. If I had additional funds and/or time, I would have bought a larger sprocket for the electric motor and a smaller sprocket for the axle. After several trials, my top speed I reached was about 30 mph.

I predicted that if I had a different gear ratio, my speed would have been even higher at maximum speed of the motor sprocket.

Figure 28. Motor, sprockets, and chain assembly

j. Purchasing the solar panels

The solar panels were purchased from Harbor Freight Tools after searching online (solar panels were much more expensive and harder to find in 2004 online). Each of the three panels on this electric go-kart prototype were rated at 20 watts and 12 volts. The price of solar panels has dropped significantly since I purchased the solar panels over 11 years ago. In 2004, I paid over $6/watt when I purchased these three panels from Harbor Freight Tools. The efficiency of the panels has also increased substantially since 2004. Figure 30 shows the new Harbor Freight brand solar panels mounted to the PVC structure that was designed. The solar panels had metal tabs with screw holes on the side where I could easily drill a screw into the PVC. Some of the solar panels that can be purchased at the time of publishing of this book at Harbor Freight do not have these tabs any longer, so these tabs will have to be added or an L-bracket will need to be modified for mounting to allow the panels to mount on the PVC.

k. Creating a mounting structure for the solar panels and installing the panels

Before building the PVC top structure to support the solar panels, a concept sketch was made in Microsoft Paint. Using this simple program, a side view profile was designed by me to give me a better idea of what the electric go-kart would look like with a PVC mounting structure. This design took much thought but after a few days of sketches, it was settled to build the PVC top using two sections of PVC, with the PVC being combined by PVC t-shaped connectors at the top.

Here is a scan of the quick sketch that was made in Microsoft Paint:

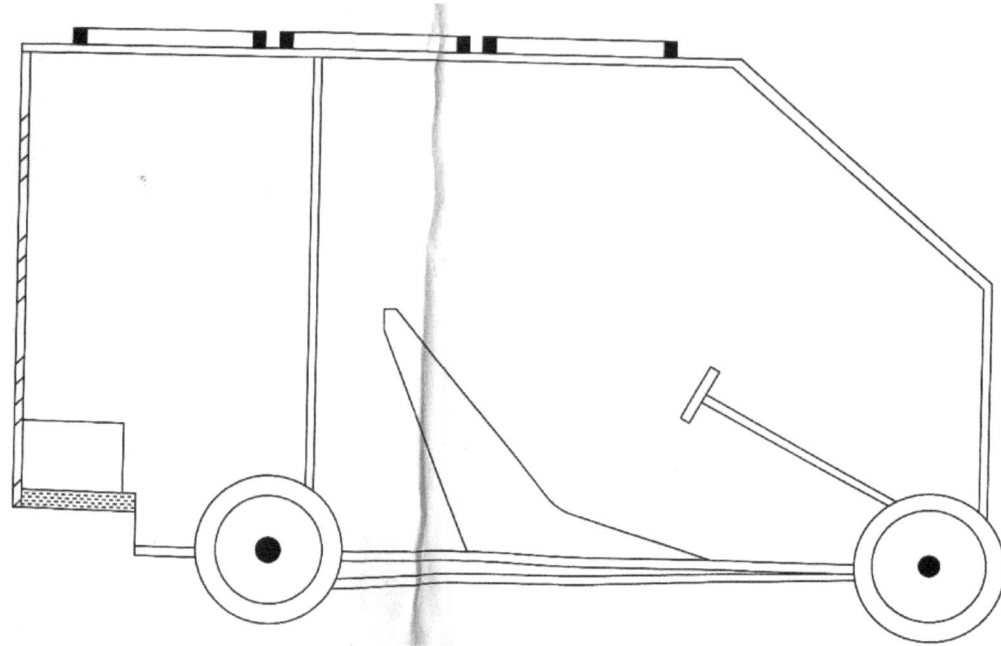

Figure 29. Microsoft Paint design concept showing the side of the go-kart and an idea how the PVC structure should look when built (scanned from original)

This PVC structure was NOT designed as any kind of roll cage, but as a temporary and cheap structure just to mount the solar panels. I decided a few weeks after I built the electric go-kart to take the solar panels off and paint the PVC the same color red as the main go-kart frame. If I were to build another version of this electric go-kart, I would actually build a metal frame and weld it to the frame. Using metal and welding would be much more cost prohibitive than using PVC, however, using metal in future electric go-karts would create a permanent solar panel mounting solution. Figures 30 and 31 have a better view of the solar panels mounted on to the PVC structure.

Figure 30. Solar panels mounted to the PVC structure.

Figure 31. The PVC structure was mounted with the solar panels

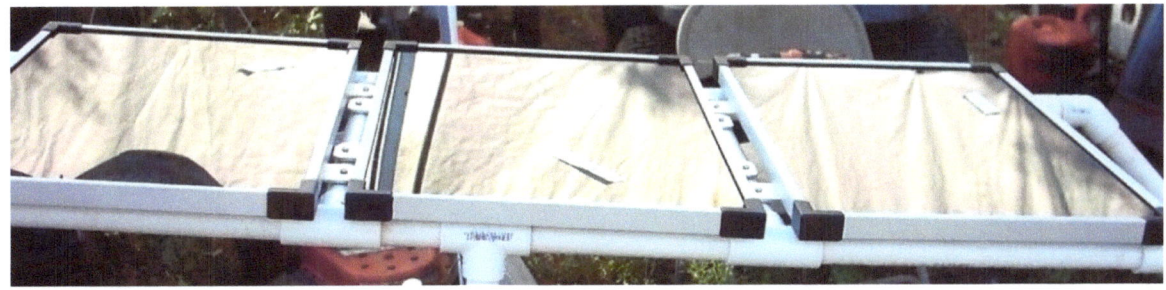

Figure 32. Solar panels mounted out of the box on to the PVC structure. The purchase stickers are still on the panels in this picture

k. Installing a front panel for the multimeters

As mentioned earlier in this book, the front panel (or dashboard) was installed with the multimeter display to monitor specific voltages and amperages. Figure 33 shows the beginning phases as I was experimenting with how to properly mount the dashboard on the go-kart. The full dashboard display can be clearly seen in more pictures later in the book.

The yellow multimeters used in Figure 34 and 35 were purchased at Harbor Freight (the local hardware store). As of December 2017, similar multimeters are sold by Harbor Freight at a price of around $4.99 or sometimes on sale for $2.99. Occasionally, Harbor Freight will offer a coupon that when you make a purchase, the multimeter is a "free gift". The current multimeters are now usually a red color at Harbor Freight and less than $5.

When building the electric go-kart, it is useful to know what the voltages and the operating amperages of the system are. One multimeter was used for the total voltage of both parallel circuits at 36 volts. This reading would tell me how charged the batteries were in the parallel circuits (the 3 batteries underneath my feet and the 3 batteries in the back of the go-kart). The 2nd multimeter was the amperage being monitored to the electric motor, thus showing how much the current draw was from the driver pushing the accelerator pedal. The 3rd multimeter monitored the voltage to the Etek motor, which would change depending on the driver pushing the accelerator pedal (or also known as the potentiometer).

Figure 34 shows a metal junction box directly below the multimeters that I used to wire into each of the circuits I was monitoring with each multimeter. The junction box was very cheap at less than $5 and kept the dashboard looking "clean" and was painted red to match the rest of the dashboard. I used a flat board to mount the multimeters to the board by taking apart the plastic casing and using 4 screws to mount the multimeters to the board.

Figure 33. Front panel installed on PVC for the multimeters to monitor voltage, amperage, and information from the solar panels

Figure 34. $2 multimeters installed on the dashboard with a junction box

Figure 35. Closer view of the multimeters installed on the dashboard

Limitations of the go-kart

The seating was slightly cramped on this conversion go-kart kit as seen in Figure 36. If a larger go-kart can be found, it is highly recommended to purchase a larger frame. When the driver puts their feet on the pedals, the room on the go-kart frame was less cramped. The driver's weight also helps with the center of gravity on the electric go-kart, which will be different for every frame.

Figure 36. The author showing the size of the go-kart. The frame was slightly small for me.

The solar panels were very easily mounted and wired in series. It was later decided to remove the solar panels and paint the traditionally white PVC piping the color red to match the rest of the frame. The solar panels were then re-installed. The biggest issue with the solar panels was that they would need to be constantly cleaned due to the pollen in the air during springtime in Florida. To maximize efficiency of the solar panels, it is always recommended to clean the solar panels at least once a week so that dirt (or in my case pollen) does not block sunlight.

Figure 37. Initial wiring to the motor, controller, and batteries

The initial connection of all of the wiring had to be followed closely, and several modifications were required due to the lack of wiring that I had at the time of building the electric go-kart. There was some creativity on how the wiring was connected based on the available sizing (some leftover wiring from other projects was utilized to minimize costs). The AXE controller had a manual that showed all of the connections and where exactly to connect the battery circuits, the Etek electric motor, and potentiometer. The yellow wires seen in Figure 37 are from the battery circuit (three of the 12-volt batteries totally 36 volts) and a separate circuit for the Etek electric motor. The spring is seen is the lower right-hand corner of the picture that shows the potentiometer circuit that is connected to the accelerator pedal or the "gas" pedal. As the driver pushes his/her foot forward on the pedal, the spring is pulled further, thus moving the potentiometer.

Figure 38. Another view of the controller wiring, battery wiring, and Etek motor wiring

Figure 39. Battery holder bar is installed along with adjuster for the electric motor

A steel bar was placed across the back batteries to help secure and hold down the batteries. This worked extremely well and is highly recommended for taking the go-kart to high speeds and turns. Without this steel bar and the two long screws, the batteries do shake around while driving. Figure 39 illustrates how secure the batteries are. Also, noticeable in Figure 39 is the additional steel flexible strap that was used to secure the Etek electric motor and to quickly adjust the tension. The steel flexible strap acted as a tension strap for the motor. To keep the chain tight, the tension could easily be adjusted with the long screw and steel flexible strap.

Figure 40. An additional three 12 volt batteries were added to the 36-volt circuit in parallel for reserve power

Figure 40 shows the 12 volt batteries that were placed underneath the steering column to give more reserve capacity to the vehicle, but also to balance the go-kart since most of the weight was towards the back. The brake link bar was also installed to the right pedal for braking.

A steel bar was also added to keep the batteries anchored to the frame. Two wooden blocks were also drilled into the metal go-kart frame to help secure the 3 batteries under the steering column. Black wire coverings were used on the side go-kart (seen to the bottom edge of the driver's seat and directly below the front batteries by the steering column). These wire coverings helped hide and keep the multimeter as well as potentiometer wiring hidden.

Figure 41. Top view of the battery and electric motor assembly

Figure 3. Completed first prototype with second 36-volt array of batteries installed

Figure 43. Side view of finished prototype.

The final prototype, with the final wiring and paint job, is seen in Figure 42, 43, and 44. The black wrap around the solar panel wire can be seen on the back brace of the PVC frame. This wrap was a black Velcro that was used to keep the wiring from the solar panels secured to the PVC frame. This Velcro wrap was purchased at Harbor Freight in a pack of about 5 or 10 for a few dollars. The Velcro wrap is not required but was easy to use and adjust as I made modifications to the go-kart. A cheaper alternative would be to use small plastic zip ties, which are much cheaper to use.

Figure 44. The author on the first trial run of the finished and painted prototype

Off-road tests were conducted in the electric go-kart after completion. The go-kart was very durable and even though it was electric, it was quick enough and fast enough to probably keep up with an equivalent sized 3 to 5 HP gasoline motor go-kart to around 30 mph on controlled conditions. In the beginning trials, the chain that was connected to the sprockets did not have enough tension and would slip. Other than the chain disconnecting a few times, the electric go-kart performed very well and had great off-road capability on dry ground.

It is NOT recommended to bring any kind of electric go-kart near water or rain due to the exposed electrical circuits and batteries. Several times in the months that followed during testing, I would bring the go-kart back to the garage when the weather turned or there were puddles on the ground. In future designs of this electric go-kart concept, I would recommend encasing all components and sealing from water to become water resistant.

Figure 45. Adjusting the multimeter that measured voltage

Figure 464. Front view of the finished solar electric go-kart.

Figure 47. Always wear a helmet when testing a prototype vehicle.

Figure 48. Side view of the solar electric go-kart.

Figure 49. Front side view of the solar electric go-kart

Figure 50. View of the batteries, controller, electric motor, and wiring

Figure 61. Back view of the solar electric go-kart

Figure 52. Giving a thumb up to the camera after a successful test run in the Florida sun

Materials List

Below are the materials scanned from the Science Fair report that was written in 2004:

MATERIALS LIST

ITEM	DESCRIPTION
GO KART KIT	FRAME W/ 5" ALUM. WHEELS; BRAKES; SEAT; STEER SHAFT
PVC PIPING	3/4" SCHEDULE 40 PIPING
3 SOLAR PANELS	12 VOLT, 450 mA, 5 Watt
MOTOR WITH CONTROLLER	ELECTRIC MOTOR WITH OPERATING CONTROLLER
6 DEEP CYCLE BATTERIES	BATTERIES TO OPERATE THE ELECTRIC MOTOR
VOLTMETER/AMMETER	METERS TO MONITOR VOLTAGE/AMPERAGE OF SYSTEM
ELECTRICAL SUPPLIES	WIRING, ELECTRICAL TAPE, LUGS

Figure 53. Official materials list (not including the wood) used for the mounting of electronic equipment

Lessons Learned

Several lessons were learned after building and testing the 1st prototype solar electric go-kart.

1. A stronger, more stable "roll cage" must be built. The PVC used as the solar panel mount was flimsy, swayed at high speeds, and was very unsafe in the event of a roll over. A stronger, more rigid material must be used for future prototypes such as metal or wood for holding more weight. This will allow for a more rigid structure and mounting more solar panels.
2. A larger frame must be used in the future. The frame for this 1st prototype was small and was almost at maximum capacity in terms of weight. There was limited room for the batteries and solar panels on this go-kart frame. The driver was also slightly uncomfortable due to the limited leg room. A 2nd prototype was initially sketched on paper in 2005 for a 2 person go-kart with cargo/storage room in the rear for batteries and equipment.
3. Additional batteries must be purchased and installed on the next prototype. As of December 2017, it is much easier to purchase batteries that can be used. Lithium ion batteries, a different type of battery, is now available for purchase. The batteries are the main factor that limited the driving time of the go-kart. More batteries generally mean more driving time up to a certain weight limit. With electric vehicles, there is a "law of diminishing returns" where if too much battery weight is added, the electric vehicle uses too much energy to move. A majority of the weight of a typical electric vehicle can come from the batteries alone. The go-kart frame was very light and the Etek motor was very powerful, so additional battery weight should not have been a big issue.
4. Adding more solar panels will help with the recharging capability of the go-kart. A better solution for the go-kart in the future would be to add more solar panels on the top of the vehicle, making recharging faster. The three small solar panels on the top of the electric go-kart did help with the recharging of the batteries; however, adding many more solar panels would have helped with the charging time. Another idea would be creating a ground-mounted solar panel charging station that is equal to several hundred watts to

several kilowatts (kW) of solar panels. The go-kart's charging plug could plug directly into the solar panel system.

5. A more aerodynamic structure could be built that integrates the solar panels into the frame. On this 1st prototype, the go-kart was not aerodynamic due to the PVC structure as well as the solid wood boards at the front for the multimeters. In a future prototype, the front end of the go-kart could be slightly curved or bent to allow for better airflow. This new design would also be more aesthetically pleasing with a curved shape for the front of the go-kart. The solar panels could be embedded into the top of a roll-cage or roof of the go-kart so that the sides of the panels do not impede the airflow.

6. A covering should be built for all future prototypes of the sensitive electronic equipment. This prototype technically could not be water resistant or water proof due to the electronics such as the exposed motor, controller, and batteries. A plastic case or fiberglass shell could be built around the electronics, protecting them from the weather elements. This go-kart was always parked in a garage when done testing, protecting it from the elements.

7. The future prototype could have four-wheel drive, which would allow the go-kart to have more off-road capability. This electric go-kart would often get stuck in light mud when testing off-road capability. Due to the electronics not being fully protected, it was not advised to handle this particular design or prototype in wet or heavy off-road conditions. A future prototype could have four-wheel drive capability and would require a new type of transmission in order to accomplish four-wheel drive.

8. The electronics dashboard could be upgraded now to just a tablet computer mounted in the front of the vehicle. This could be integrated with several digital multimeters or using wireless/Bluetooth to communicate with the tablet. Today's technology (as of December 2017) allows for wireless communication of electronic controllers, motors, solar panels, etc. Instead of using cheap $3 or free multimeters, the driver could have a touch-screen available to them that could display all the relevant data from the voltage/amperages of the electric motor, batteries, and solar panels. A user interface (UI) could be custom built for the driver of the vehicle. A custom reading for showing the "gas gauge" or how much

energy was left in the batteries could be created to help the driver know when the batteries need to be recharged.

9. Lights would need to be added to future go-karts. This prototype did not have front or rear lights or even brake lights. These lights could easily be installed on future prototypes with a switch system integrated into the battery system. Light switch toggles could be added to the dashboard.

YouTube Videos

A complete YouTube playlist was created with videos of this electric vehicle for your enjoyment and knowledge. This YouTube video playlist is available at https://www.youtube.com/playlist?list=PLEVfP2SpIQhcD5Pv9ojnJPGNMmeBoTR1z Several videos are included in this playlist such as "My Solar Powered Go-kart" and "DIY Wind Turbine and Solar Electric Car".

I usually respond to YouTube comments within a week or less. If you have questions, please feel free to comment on my videos of the electric go-kart or send my YouTube channel a message at www.youtube.com/nuclearboy2003. I do read all comments and hope that you can build your own DIY solar electric vehicle one day.

About the Author

Eric A. Layton received a Mechanical and Aerospace Engineering degree from the University of Florida with an extensive understanding of solar power, bioenergy, and sustainable solutions. Mr. Layton has specific skill sets in designing solar systems, design and engineering of biofuel systems, and nuclear reactor training from the University of Florida. Mr. Layton has a proven ability to use applied research, especially when he designed and built the largest educational solar powered biodiesel facility in the United States in Gainesville, Florida. He was also the lead researcher into using solar power to grow algae in order to produce biodiesel. Mr. Layton grew several algal strains for future algae biodiesel research including construction of a 100% solar powered photobioreactor system while at University of Florida. His current endeavors involve consulting for solar electric systems as well as solar hot water systems in the United States.

www.ingramcontent.com/pod-product-compliance
Lightning Source LLC
Chambersburg PA
CBHW041932240526
45473CB00034B/926